AF880788

5 Steps to a World Class Maintenance Organization

By The Maintenance Geek

Robert J. Kravontka

ISBN-13: 978-0692683910

ISBN-10: 0692683917

DEDICATION

This book is dedicated to my brother John Kravontka president of TPM Unlimited, one of the great maintenance professionals of the world, who has forgotten more about maintenance than most will ever know.

Introduction

What is it costing your company being in Break Down Mode?

Billions of Dollars Nationwide... Find out here how to get out of the break down mode and approach World Class

Robert Kravontka is a maintenance geek, who was tired of living break down to break down. He has studied the thought leaders in the industry, and is bringing that knowledge to his followers at www.maintenancegeek.com

Are your best maintenance professionals ready to retire? Because most of the best apprentice programs were halted for budgetary reasons many years ago, the best trained maintenance personnel are reaching retirement age soon. Some of these maintenance professionals did get informal training over the years, there are still some really great ones out there.

Do your younger people need to work smarter not harder? With some good training in the basics you can make some big gains in speed and efficiency, hence lower costs to keep your most critical equipment running.

Lots of companies are up against these issues. A big portion of smaller manufacturers are functioning with one maintenance person for the entire facility, and they only do repairs when equipment breaks down. $1.00 invested in their training will return $10.00 in less downtime and breakdowns when your most critical equipment needs to produce the most.

With the efforts of lean manufacturing putting equipment in a cell formation to produce only what you need when you need it, working toward the shortest lead time, a couple of things have changed. Now if a machine goes down in a cell, the entire cell is down, and you are effectively out of business. In the old days you just moved production from one machine to another in the press or turning department. Downtime is much more visible now and more costly because of the disruption it causes.

What does the lack of good maintenance cost worldwide?

Billions!

When government cuts back on road maintenance by 5% to fit into the budget, what grows typically are pot holes, lots of them.
This leads to broken springs, flat tires, bent rims, alignment issues, which equals significant costs for users of the roads, who saved very little in reduced taxes, but exponentially paid for the results of poor maintenance.

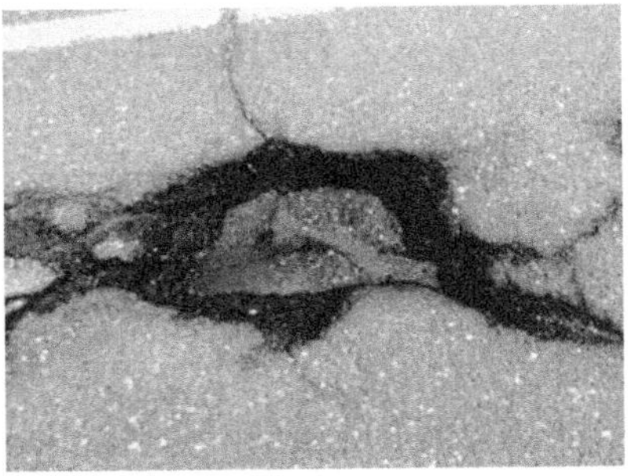

When a piece of manufacturing equipment goes down, the losses start with immediate production for sure, the operator must be sent somewhere else to work, while the repairs are being made. This lost production will generally need to be produced on overtime, in today's lean environment the machines downstream in the process will stop producing because of lack of parts. The critical repair parts may need to be flown in overnight if they are available, and if not on the shelf. Maybe you will need to fabricate new parts for older machines that are no longer available.

In General breakdowns are 5 to 10 times more expensive that preventing it in the first place.

In the 70s and 80s audible sensor were put in automotive brakes. These make a squealing noise when the brake pads get worn to a replacement level. This is a condition based PM.

My wife didn't like the noise so she kept turning the radio louder to drown it out.

By the time I got to drive the car, not only did I have to replace the pads, but the rotors, and calipers. One of the brake lines had started to overheat, so the brake fluid boiled , then leaked and all the fluid, had to be replaced.

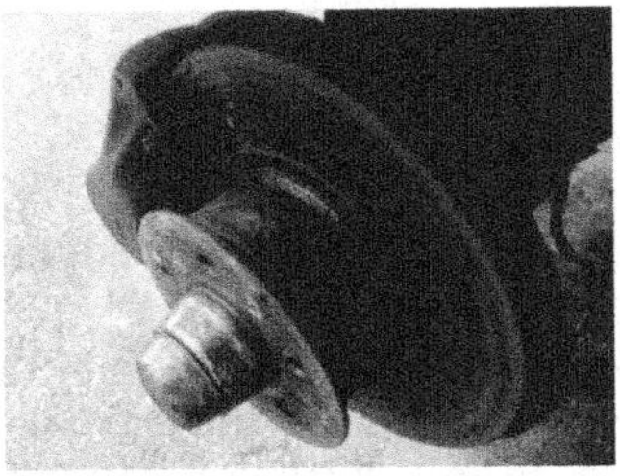

Pads would have been around $60.00.

$850.00 later the audible sensors were again quiet.

Breakdowns are a minimum of 5 Times or more expensive than a properly timed preventive maintenance schedule.

We are going to look at how a small company with stretched resources can use these principles to improve the working lives of their people and improve their production for very few new investment dollars. You can save money with some better maintenance practices and be able to pay for much of the training and equipment to get and stay out of the breakdown mode. Better maintenance practices not only improve production numbers and increase profits, but also reduce energy costs. The utilities tell us that every dollar in wasted energy, takes $7.00 in "sales" to make up.

CONTENTS

Step 1

The Building Blocks of World Class Maintenance

You must start with some data.

Think of it as a doctor's visit. Your doctor will not prescribe a medication or solution before he examines you and collects some data. At a minimum, blood pressure, weight, temperature, and what you think is wrong.

Should it be any different trying to diagnose your maintenance process, before you prescribe some remedies?

Here is a very simple observation form. Start with your critical pieces of equipment and find out when they are actually running and why they stop.

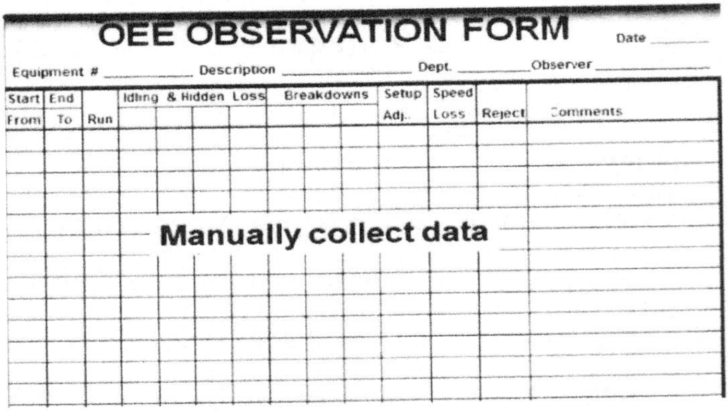
OEE OBSERVATION FORM

Date ______

Equipment # ________ Description ____________ Dept. ________ Observer ________

Start From	End To	Run	Idling & Hidden Loss				Breakdowns			Setup Adj.	Speed Loss	Reject	Comments

Use these building blocks of World Class Maintenance.

- Collect Data.

- Determine your critical pieces of equipment. Rate them in order of criticality. Only look at the top three critical pieces of equipment to start, and also identify the top two pieces of equipment that cause you the most problems. The PIA equipment that is taking up your time and resources. They will most likely be different.

- What skills are at your disposal?
  You may be surprised at some of the hidden talent you have available.

- What do you need to do to keep the equipment running vs. trouble shooting it and fixing it? Get out of the breakdown mode. Change your mindset.

- OEE (Overall Equipment Effectiveness) is more important as you start to flow your products. i.e. LEAN. If one component of the cell stops, the entire cell stops. OEE multiplies down the line as more equipment is added to a cell.

- Evaluate your maintenance department on a World Class Scale.

- Benchmark against companies in similar industries. Although it is more important to know what you are doing well, and what you need to improve upon.

- Maintenance Mapping is similar to value stream mapping in lean terms, by looking at what actions add value to keep the equipment running, and what actions add no value and should be stopped or reduced.

- What is the vision for your maintenance organization over the next 2-5 years? Look at proper staffing and training with a goal to reduce downtime by a set percentage and increase OEE on critical equipment by a determined amount.

Let's consider the doctor scenario again.

You visit and he asks you things like, how long has it been since your last visit? What was it you were here for, and has your weight and blood pressure changed?

You would say **W**hiskey **T**ango **F**oxtrot… Shouldn't the doctor know that?

Now this doctor would, if you came in unconscious, trouble shoot you, and try to get you going again. In the office visit, his goal is to keep you running and healthy. This should be the goal of a world class maintenance group, to move away from emergency lifesaving surgery and toward keeping the equipment running.

This is a change in mindset. Now you look at a machine that is down and troubleshoot it, but better you look at running machine and think about what it will take to keep it running. What spare parts should you keep on hand, should the controls be updated? Is the lube system workable and replaceable? Should the filter systems be upgraded or at least cleaned? Where could the lubrication be improved or at least topped off and how could the contamination of vital components be reduced.

There is a lot of data to collect. If the data is telling you something, use it.

Don't be like this guy.

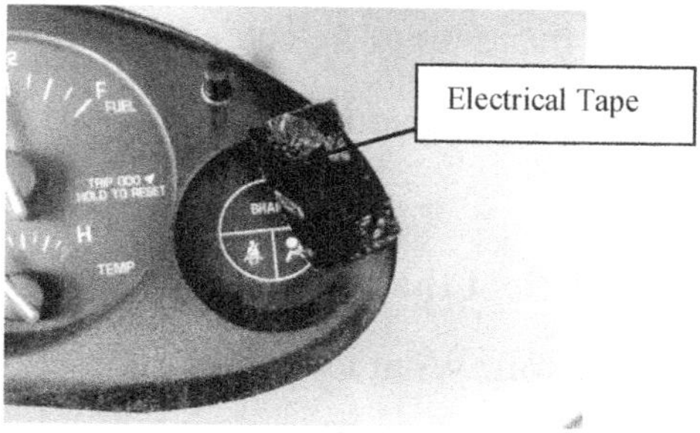

Do Something about It!

Back to collecting data, here are some questions.

Where do you collect data?

Start with your top three to five critical pieces of equipment, if they go down, big trouble in lost production and big dollars to get back on line. As your equipment and systems develop add more in a similar manner.

How do you decide what equipment is the most critical? Give your equipment a rating by asking how often is the equipment used each day? Put a 1-10 number next to it. Is there back up equipment available? Rank it. Will this failure affect other processes? Is the equipment leaking? Add a safety ranking. Total these up and decide the order your equipment needs to be addressed.

How do you collect the data?

Direct observation. What do your maintenance professionals do all day. (Be nice). Do they go from breakdown to breakdown, and occasionally when they catch their breath accomplish a couple of PM's? Maybe the new guy or the apprentice will do all the PM's and leave the old pros to put out the fires. These guys are good at it, and are rightly proud of their abilities to turn around a broken down machine.

Do you have **measuring systems** you could be using? Do you have a work order system? Can your operators put in work orders?

Is your **CMMS** up to date, if you have one? There are computerized maintenance management systems out there that will let you try them for 30 days to see if it adds value to your company. These are great, but only if you actually use them.

Do you have an **SAP** without a maintenance module? Do you know or even care what that is?

If you do **not** have any of these sophisticated measures available try something **simple and visual.**

Begin by measuring your **repetitive failures.**

Go to the equipment that is the most critical to the operation, and perform a 60 Minute OEE measurement. OEE **(Overall Equipment Effectiveness).**

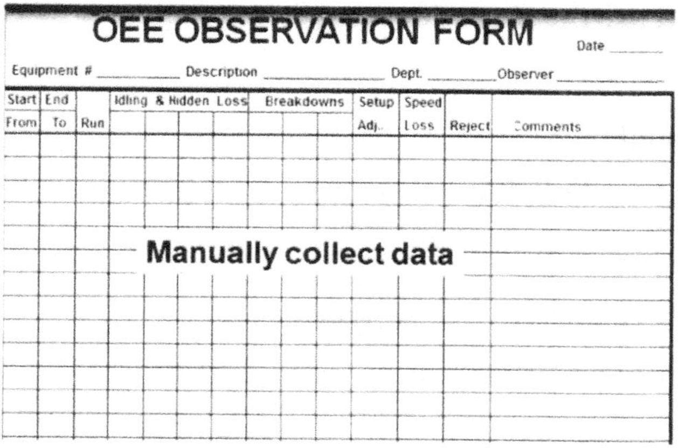
OEE OBSERVATION FORM

Date ______

Equipment # ______ Description ______ Dept. ______ Observer ______

Start From	End To	Run	Idling & Hidden Loss	Breakdowns	Setup Adj..	Speed Loss	Reject	Comments

Manually collect data

OEE is a way to measure the productivity of a piece of equipment. You are actually measuring the 6 Major losses.

A simple 60 minute OEE measurement would include the 6 major losses. You need to know what these losses are doing to your productivity. These losses can only happen when you **plan** to run the equipment, not weekends and off hours, or breaks.

What are the six major losses?

Availability

If the equipment is being set up or is broken down, you lose total availability.

Breakdowns = No availability.

Setup & Changeover = No Activity

Performance

When the equipment is idle or being repaired for minor stoppages.

Idling & Minor Stoppages

Record each time the equipment is idling, waiting for parts, or is stopped because of a jam.

Reduced Speed

Some operators run the equipment a bit slower because it runs better, but the difference between what it should be running at, and actual are hidden losses.

Rate of Quality

If the equipment requires warm up time before you get a good quality part, it's start up loss. If you make scrap it's considered a quality loss. During start up try to reduce the time to make a good part.

Quality Defects & Rework

Always look for ways to run good parts every time

By far the **largest loss category** would be Idling and Minor Stoppages, or what we call hidden losses, since in general no one but the operator can detect these failures. This data can tell you how effectively your equipment runs when you plan to run it. Identify your major losses, and then determine the equipment's current output, this will help you determine the baseline for this equipment.

When you cannot get enough data in 60 minutes to generate some reliable OEE data consider an 8 Hour OEE. Try using someone to collect the data on light duty, or a summer intern. As you start to understand your 6 major losses, on the first few machines, this will help develop a roadmap to find problems and increase capacity.

Setting targets for OEE.

Setting a world class target of 90% becomes a

corporate target, there is no ownership developed. Corporate OEE becomes a weapon not a tool to improve.

Try setting a best of the best – (BOB)… This is taken from past data, at the machine, based on the **(best availability X the best Performance X the best Quality).**

This becomes a realistic target that the operator knows the machine is capable of. The team can work toward attaining it. Once the target is attained for 4 weeks, it is time to raise the BOB. Set your goals for each category and continually improve upon the past performance.

Typical OEE measurements are in the 40% range. With minor stoppages the largest at 24%. Breakdowns at 15%, Setup 16%, and quality scrap 6%. These are typical; each piece of equipment will be different for each type of industry.

A world class OEE is in the 90% range, with minor stoppages less than 4%, breakdowns less than 3%, set ups less than 2%, and quality scrap less than 1%.

These OEE measurements can be achieved, by steady, gradual improvement.

How to convert OEE into Dollars

OEE, or Overall Equipment Effectiveness = Availability X Quality X Performance

An increase in OEE means that at least one of the 3 factors, availability, quality or performance, have been improved.

- An increase in availability represents creating more time for production.
- An increase in quality represents more good products making it through the production process.

- An increase in performance represents producing

more in the same amount of time

Every process has an output per hour.

Every product has a selling price.

Example:

Assume;

7 hours available to run per shift. **200** pieces per hour when running an **80%** quality rate and each piece sells for **$10.00**

7 hours X 200 pieces X .8 Quality Rate X $10.00 each= **$11,200.00 per 8 hour shift.**

A 10% increase in OEE means:

Either

.7hrs more available time per shift = 140 more pieces times $10.00 is **$1,400.00 per shift.**

10% better quality = 20 More good pieces per hour times $10.00 each is $200.00 more per hour or **$1,400.00** per 7 hour available shift.

A 10% increase in performance equals 20 more pieces per hour, or **$1,400.00** per 7 hour available shift.

Likely a combination of outcomes will happen.

A 10% increase in OEE is reasonable, and much more may be possible. Do you think you will be able to get more funding for your maintenance projects as you can show $1,400.00 in increased production per shift X ___shifts per year?

Gemba Walks.

Gemba is a Japanese term meaning "the real place." Japanese detectives call the crime scene *Gemba,* the idea of *Gemba* is that the problems are visible, and the best improvement ideas will come from going to the *Gemba.* The *Gemba* walk, much like Management By Walking Around (MBWA), is an activity that takes management to the front lines to look for waste, and

opportunities to practice practical shop floor improvement. Encourage your maintenance people to practice Maintenance by Walking Around or MBWA while they are walking between jobs. They need to look and listen.

What should a good Gemba walk entail?
Look for two things and Listen for one thing.

Look for contamination. Is there "Stuff" (This is an euphemism for the other four letter "S" word) getting into places it should not? Are filters visible and clean? Are there visuals on the equipment to alert the operator of a problem? Is there an operator check list with visuals for the operator to walk around and check the machine with each day?

Look for Lubrication points.

Are they color coded to reduce cross contamination? Is there a way to check if lube points are being serviced? Are gauges clean and marked for proper indication?

Are your oil levels easy to check?

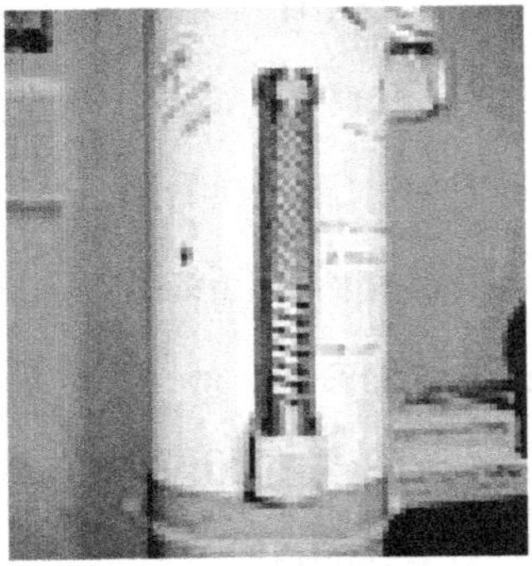

Have you put zebra cards behind them to make the level and clarity stand out?

Listen for Air leaks.

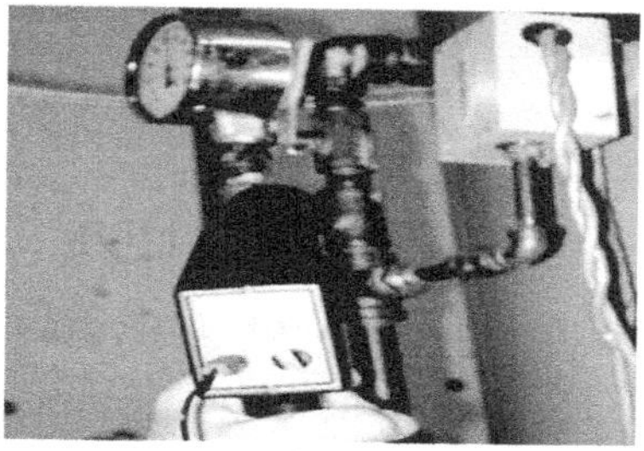

These are pure waste. Do not fix them until you can put a dollar figure on what they are costing, then fix then right away. These can help pay for some predictive equipment. Listen for bearing noises, chain clanks, and squeaks. Identify areas to put Lexan viewing ports for routine visual inspections.

Will benchmarking to other companies in your business segment help you? Maybe.

You need to know where you are, before you can figure out where you are going and how long it will take to get there. Benchmarking will help you determine where you are relative to others in your industry. It is more important to know where you are planning to go. A good gap analysis may be able to measure this for you.

Benchmarking measures your key operations and personnel, there are over 70 aspects of maintenance performance you may want to measure. Use these assessments as a road map to begin building an effective maintenance strategy, long and short term.

From this benchmarking data develop a maintenance vision. As a maintenance organization, or just one person, what do you want to achieve, keeping your machines running is a good one. Who are your customers? How big is your organization? Doing a good gap analysis on where you are with the basics and where you want to be, may be all you need to develop a good strategy.

Is your department called equipment repair, or the maintenance department? Do you maintain the equipment so it will not break, or do you and your people simply repair it when it breaks.

From this vision put together a strategy to get there.

Develop answers to these questions.

- **What are your critical pieces of equipment?** Start with the top five. Three that are critical to the operation and 2 that are just a (PIA) by taking up way too much time.

- **What data is available,** or should you collect to guide you to what to add to your PM program?

- **What skills are at your disposal?** Do you actually know what skills your people possess? Once you find out, put a plan together to acquire the skills your team lacks. There are lots of programs through your departments of labor and local universities to help, and they are often times funded.
- **How are you going to get out of breakdown mode?** The balance of this book is about just that. Every hour you spend predicting and

preventing failures, will free up 5 – 10 hours of time from the emergency repairs that eat up your time now.

- **Who else could be your eyes and ears?** Your operators are at the machines all day, no one knows more about the equipment than they do. Give them some visuals so they can predict a failure and get the news to you beforehand.

- **What could you ask them to look for and listen for?** Some simple things like marking gauges and oil levels so they can keep an eye on them. Maybe some filter material over a motor or electrical cabinet to monitor when it should be changed. If you put these filters on with Velcro, maybe they can change it, and free up more your maintenance's time.

Use the PF curve for guidance in mapping your journey out of breakdown mode.

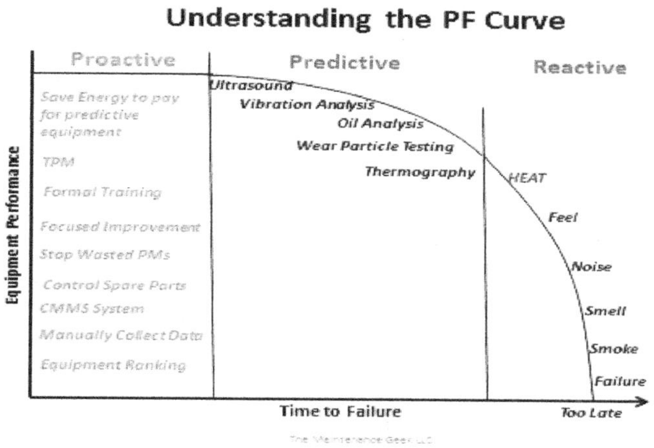

The **P-F** (Potential to Functional Failure) curve is based on using the correct technology and tools to detect possible failure. This enables a window of time for repair or replacement. The X axis reflects time from failure start to machine failure.

On the right side of the curve, when you smell something wrong, or feel vibration or see smoke it is typically way to late, and you are in breakdown mode. Our goal will be to move left toward some predictive tools and then some proactive actions to develop a good PM program to pull you out of this breakdown mode.

STEP 2

Control your Spare Parts

This is the number one reason for maintenance inefficiencies.

One customer had 16 million dollars' worth of spare parts, which may be the right amount, but after they went through them it turns out 8 million dollars of them were for equipment they no longer had!

The equipment manufacturers can be helpful with a list of recommended spare parts, although with some data collection, you will want to hold only the parts that meet the 24 hour rule. If it takes less than 24 hours to get the part, don't stock it. Only stock mission critical parts. Ask yourself these questions, Are these parts critical to keeping up with customer demand? They cannot be obtained within 24 hours, or their cost has no impact. **You must keep these in stock.** The reality is that you do not always get to make that decision.

The Materials Management and Purchasing functions are often pitted against Maintenance, Facilities, Engineering, and Operations with respect to; the number and types of stock keeping units or (SKUs), capital and critical spares, obsolete inventory, inventory investment and turnover. Program operation of stock-outs, and the perception of inventory as "cheap Insurance," tied to the lack of knowledge with respect to the total cost of holding inventory. "Cheap insurance" is based on the thought process that inventory is typically expensed when received vs. issued and that cost is small compared to any potential loss of revenue that could occur if operations/production was halted for

any measurable period of time due to not having the replacement parts.
Spare parts management is a science: It is not unusual to see an average of 20-30% of maintenance crafts people's time to be used for finding parts and acquiring material.

Using an ABC analysis tells us that inventories of an organization are not of equal value. Thus, the inventory is grouped into three categories (A, B, and C) in order of their estimated importance. Primarily, a spare part inventory is generally analyzed through a failure mode effects and criticality analysis or FMECA to obtain a first reduction of the problem, meaning that only the most critical spares are analyzed. Thereafter, an ABC analysis according to Pareto's principle of the most critical spares is carried out to classify the items. In this way a second reduction of the items is obtained (i.e. only the spare parts classified into the most critical class will be considered in the next step). The conventional ABC analysis adopted by many firms, classifies the plant components into three classes of criticality: very important (A-class), important (B-class), and less important (C-class); only the components belonging to class A require management and maintenance attention. An accurate re-definition of the stock levels for the different spare parts should be taken; thousands of items may be potentially held in inventory by a typical company, but only a small portion of them deserve your close attention and accurate control

Besides being the central hub for maintenance, the storeroom also provides functions that are absolutely critical to the maintenance operation. If the storeroom is run improperly (such as poor inventory accuracy, parts unavailable when needed due to poor replenishment practices, the rest of the maintenance operation will not achieve high service levels of equipment availability and reliability.
The Storeroom as a Service Provider - If you follow the flow of technicians and materials in a typical maintenance operation, a common pattern emerges.

In general, you have a central hub and the spokes of a wheel - where the hub is the storeroom, and the spokes flow out to the equipment. This places the storeroom in the position of a service provider, with the rest of the maintenance organization and manufacturing operations as the customers. Perhaps one of the main ingredients of success for a service provider, besides delivering what is promised, is to manage expectations.

Develop a clear spare parts stocking criteria

To stock or not to stock. One of the flaws in most spare parts management systems is the absence of clear rules for when to stock an item.

The absence of any guidelines from maintenance and management forces the storeroom into a process of poor decision making. The result of this is that they stock items that don't require stocking and don't stock items (sometimes critical items) that they should.

Step one is to determine a series of stock categories

that enable clear separation of stock types. A, B, or C, then quantities of each.

Step two is to use a series of structured questions to explore alternatives. For example:

1. Is the item obsolete or likely to be obsolete in 6 months?
2. Can we use something else that we have in stock (substitution)?
3. Can usage be planned within the supply lead time (Can you forecast usage, using predictive tools)?
4. When required can the item be delivered by a vendor in an acceptable lead time (quick/local delivery, overnight)?
5. Can the asset in service be repaired in an acceptable lead time (repair rather than replace)?
6. Is the item being ordered for a project or upcoming PM (not stocked)?
7. Can the need for the item be eliminated (engineered out)?

8. Can we use someone else's stock (share/pooling)?

Provide Clear Guidelines on How Many Parts to Stock. Once you have established clear spare parts stocking rules the next step is to establish clear guidelines on how many to stock. By developing and implementing clear stocking guidelines you provide the logic for decision making on stock levels and calculating reorder quantities. The advantage of this is that it provides guidance to your team, continuity for decision making, and provides the basis for future audits of inventory holdings. The principles of lean and Kanban can be very useful here.

Accept that Some Stock Outs are OK. One of the main fears of most reliability and maintenance engineers is that, at the time of actual need, the required spare part will not be available. This is commonly referred to as a stock out. But how do you eliminate all stock outs? The short answer is that you can't, not without tying up massive amounts of money that would be better used elsewhere. Even then, you could not guarantee eliminating all stock outs. This would possibly require holding multiple units of every part in your plant. That's not reasonable. What you can do is accept stock outs in areas where they have little or no impact or where a good supply alternative exists. Then work to eliminate stock outs for the items with high criticality. There are essentially two causes of stock outs. The stocking guidelines are not correct for the current level of demand. Or, there is some delay in the supply chain, including your internal processes associated with purchasing. When measuring stock outs it is important to differentiate between these different causes. This is because adjusting your stock levels (usually upwards) to account for a temporary supply event may not be the best course of action. One quick and easy way to collect a record of stock outs and enable follow up to root cause is to use a clipboard. At most locations the stock out rate is unlikely to be so high that you would need any automated solutions to collect data, so a good clipboard works well.

All you need is a form with headings such as part number, description, date, person requesting the part, and likely cause of the stock out and you will instantly create a record and get it in real time.

Storeroom Security.

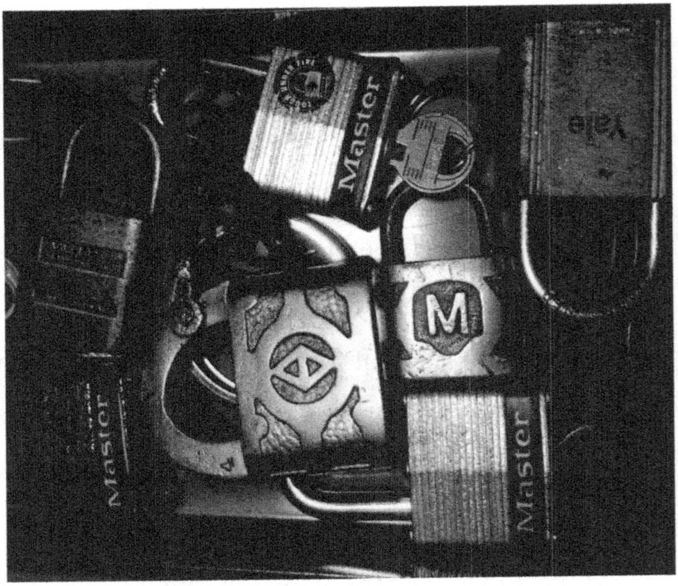

Many companies are fully aware of the importance of maintaining security around their spare parts inventories. They secure their spares in lock-up areas, ensure staff are on hand to manage the level of spares requests during periods of high maintenance activity, and provide a solid approach for identifying storeroom entrants during the afternoon and night shifts. Try not to leave any back doors open. Removing items from the storeroom without proper record keeping, no matter how honest the intent, will not trigger reordering when required and next time you need that item it will not be in stock.

If you think about the cause of inaccurate stock level data you quickly realize that there really are only a few, and two of the main ones are:

1. Items being taken (or picked) without being recorded.

2. Items being placed into the store, either from a new delivery or via a return to store room, without being recorded. By not paying attention to

storeroom security, companies effectively sabotage their spare parts efforts.

World Class Maintenance meets lean and 5S.

Use set up carts with tools located in sequence of use, and on shadow boards so missing tools can be identified.

Time is money, and lost time is lost money.

Use SMED (Single Minute Exchange of Dies) techniques to standardize tools and fasteners. – Quick Connect Clamps – and Color Coded Pneumatic Tubing.

Tool organization reduces lost time and money.
More time is spent each day looking for tools than on actual wrench time. **5S** your tool boxes and note the improvement in response time for repairs. Monitor these improvements and require a periodic tool box inspection. Think you will get some push back? Sure, but continue to discuss how important becoming world class is to your group.

How quickly can you find the tool you need in this tool box?

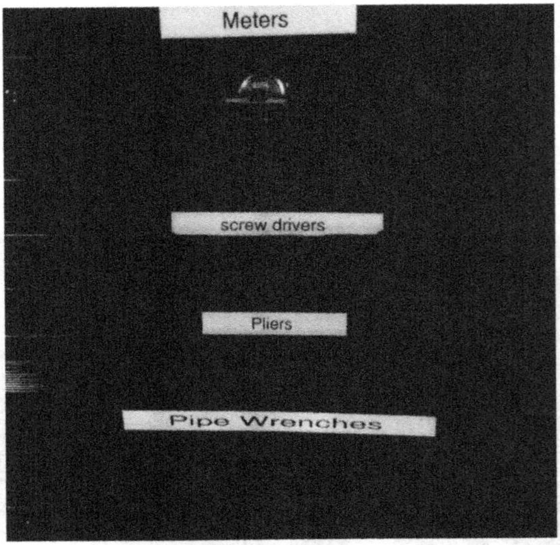

How about
now?

STEP 3

STOP WASTED PM'S

What are "effective" PM's?

Typically 50% of PM's are wasted. An example would be doing a PM in an electrical panel, the goal is to tighten the 100 or so connections, typically 1 or 2 will need tightening. This is mind numbing work, and can lead to damaged connectors. Unintended consequences. If predictive tools were used, like an infra-red camera, only the 1 or 2 connections would need to be tightened, and the broken connector would have been
avoided.

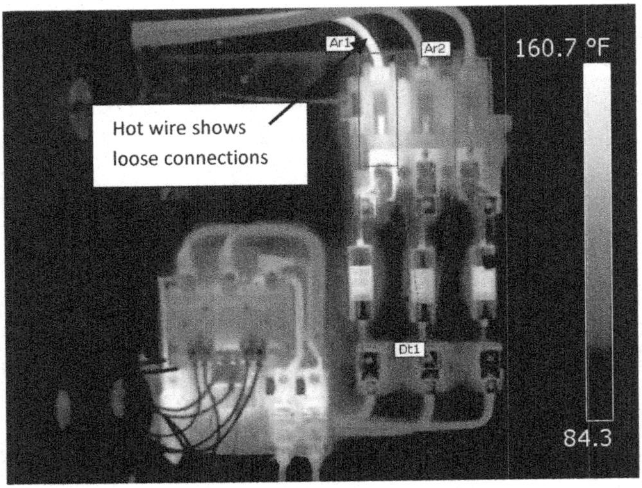

Time vs. Condition PM's.

Every 2000 hours according to the manual or by doing an oil Analysis first may reduce frequencies of oil changes.

Develop route based oil analysis to trigger oil changes, instead of just changing oil on prescribed time intervals.

Condition-Based Maintenance (CBM) is a maintenance philosophy used to actively manage the health of equipment in order to perform maintenance 'when" it is needed. CBM can drastically reduce operating costs and increase the safety of equipment, requiring maintenance. Reactive maintenance can have severe performance costs. Preventive/predictive can be a cost savings move vs. scheduled maintenance which may replace parts before the end of their useful life. CBM optimizes the tradeoff between maintenance costs and performance costs by increasing availability and reliability while eliminating unnecessary maintenance activities.

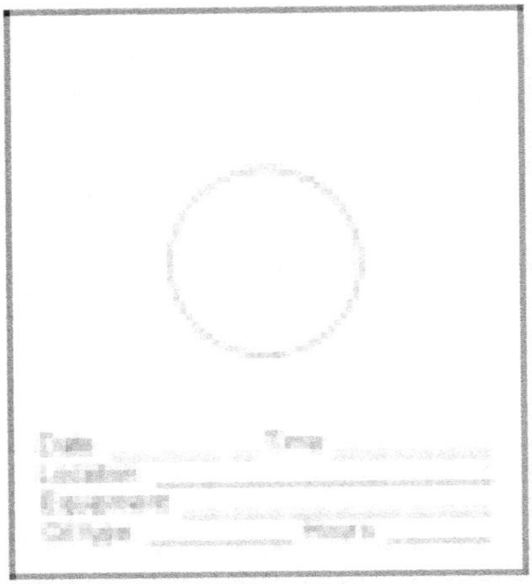

Oil Test Card ready to take a drop of oil. Knowing where to take the sample from is very important, ask your lab people for some training on this.

Developments in recent years have allowed extensive instrumentation of equipment, and together with better tools for analyzing condition data, your maintenance personnel are more able to decide, what is the right time to perform maintenance on some of equipment? Ideally condition-based maintenance will allow your maintenance personnel to do only the right things, minimizing spare parts cost, system downtime and time spent on maintenance.

First and most important of all, the initial cost of a complete condition based monitoring system can be very high as it requires improved instrumentation of the equipment. Often the cost of instruments can be quite large, especially on equipment that is already installed. Wireless systems have somewhat reduced the initial cost. It is important to decide the importance of the investment before adding Condition Based Monitoring to any of your equipment.

Secondly, introducing condition based monitoring will invoke a major change in how maintenance is performed, and potentially to the whole maintenance organization in your company. Organizational changes are in general always difficult, also the technical side of it is not always very simple. Even if some types of equipment can easily be observed by measuring simple values as vibration (displacement or acceleration), temperature or pressure, it is not easy to turn this data into reality about the health of the equipment. This situation can be eased by, using predictive equipment in the PM process.

As instrumentation and information systems tend to become cheaper and more reliable, CBM becomes an important tool for running a plant. Better operations will lead to lower production cost and lower use of resources. Lower use of resources may be one of the most important items in a future where environmental issues become more important.

A more down to earth scenario where value can be created is by monitoring the health of your car motor. Rather than changing parts at time intervals, the car itself can tell you when something needs to be changed based on cheap and simple instrumentation.

Changing hydraulic oil can also be likened to looking into the sink after you shave, and seeing the residue stuck to the surface of the sink. Consider a flush and change to get these contaminants out of the system, and thereby extending your oil change intervals.

Looking back at the P-F curve, predictive equipment can help you plan your repairs before the equipment starts to fail.

Predictive Equipment:

Ultra Sound for air leaks, proper bearing lubrication, and steam trap health.

Infra-Red for motor life, any motor over 150 degrees needs to be watched, motor life drops off exponentially with each 20 degrees of temperature increase. Loose electrical connections, are a safety and fire hazard. Process anomalies like furnace insulation, along with facility monitoring like roofs and air infiltration can pay for the equipment with lower energy costs.

Oil Analysis to extend intervals between changes. Accomplish this by a route based oil testing program.

Vibration Analysis to predict a host of motor, connector, and bearing issues.

Operators can be very predictive. They must own their machines and communicate the health of their machines to maintenance.

This picture is a Zebra card that shows oil level. This also shows quality, as it is not milky or speckled with contaminants, or discolored from heat.

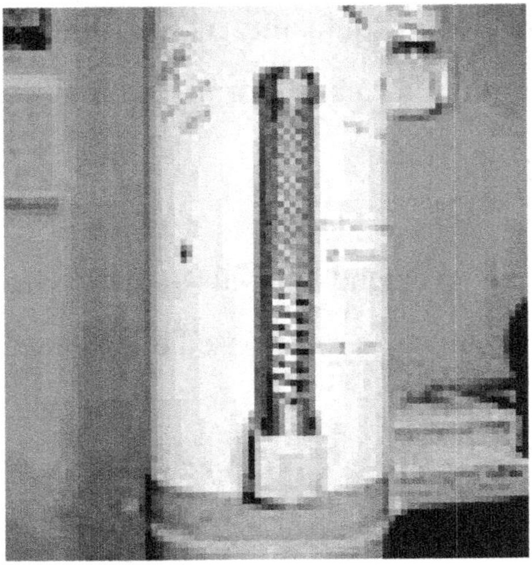

Make checks visual for contamination, in your air oil and water filter systems.

Visual and accessible – Visual for identification of wear on belts, chains and accessible for maintenance to repair them quickly without huge disassembly or LOTO. (Install Lexan covers).

Using PM sign off sheets with numbers and visuals have shown typical results from 2 people for 8 hours to 1 person for 2 hours.

Talk about wasted PM's…

Measure and track your exhaust air velocities with a velometer.

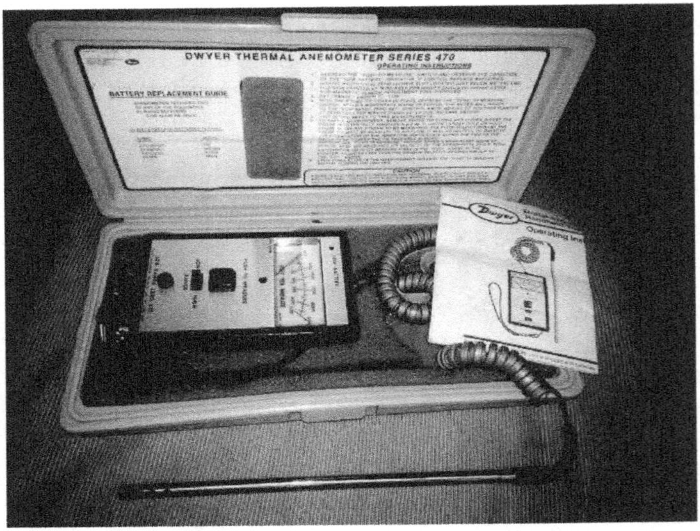

Test and track a few locations… Huge amounts of contamination grow as venting systems start to clog. Prioritize maintenance in these areas. Most companies never look into these areas and pay a big price.

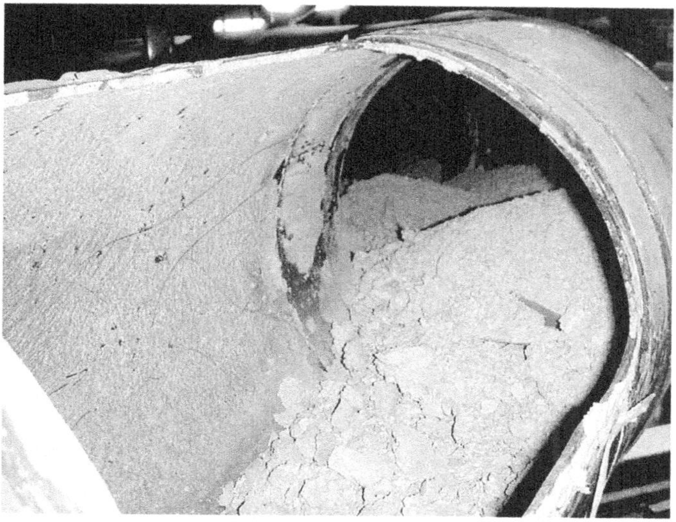

Do something about it!

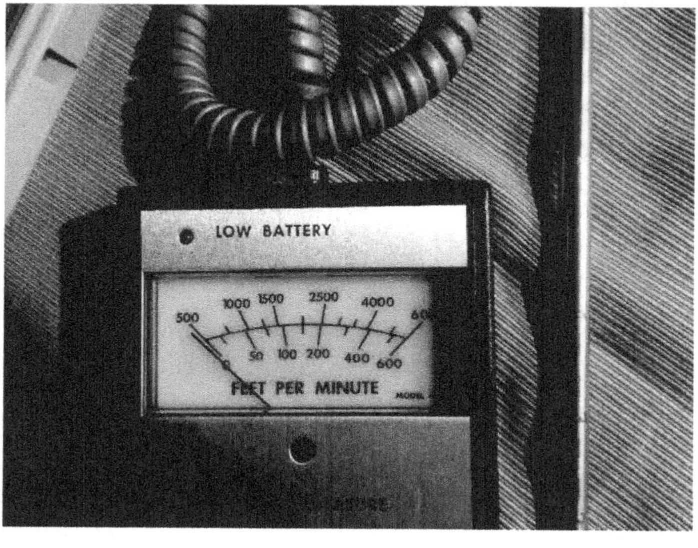

Change your PMs from being time based to condition based. When air velocity falls below predetermined levels because of blocked ventilation, do something about it.

In addition to condition basing your oil and air related PM's, training for your oil technician can pay big dividends.

It has been called the 5 rights of lubricants:

Right Type:

Right Quality:

Right Amount:

Right Place :

Right Time:

If you use work orders for lubrication PM"S teach your lube tech to review the work order for materials and tools required to accomplish the task without walking back and forth.

Study the OEM manuals , Do you have them? Are they in the right place? Is there a place? Any help online?

- Is your lube room labelled?
- Is it compliant?
- Do you have Material Safety Data Sheets (MSDS) available?
- Can you easily see the inventory levels?
- Is the area clean and are the bungs on?
- Are your breathers and transfer containers in place and color coded?
- Are there contamination issues and is the lube room clean?
- Are your grease guns stored upright or horizontally?

- Does your lube tech have a place for work orders and do they keep your lubes identified and clean?
- Does everyone involved know the safety procedures for Lock Out Tag Out (LOTO), do they need training?
- Are your condensate buckets labelled and checked on a schedule?
- Do you schedule visual inspections of facility equipment like transformers?

Vertical storage allows the oil to migrate out of the grease. Store them horizontally.

Consider something simple, like how to use a grease gun.

"In any lubrication practice, cleanliness is paramount. Nothing wipes out the value of lubrication more effectively than contamination by abrasive grit or other foreign matter.

The area where the grease gun is filled must be clean, well lit, have a work tabletop at a convenient height and clean rags at hand. The workspace must be free of dust and atmospheric contamination.

A competent lubrication practitioner must verify the work of those being trained and periodically inspect, without prior notice, the trainee's work. The use of a checklist is helpful.

A dedicated grease gun is necessary for each type of grease being used. Powered grease guns are not desirable, except in instances where very large amounts of grease are required.

Before filling the grease gun, it must be thoroughly disassembled and inspected under good light for harmful debris inside and out. If there is any doubt as to the quality or cleanliness of the grease gun, procure a better quality one. A pistol grip grease gun is recommended as it allows one-handed operation, with the other hand being used to hold the grease delivery hose or adapter.

Using a grease cartridge is the preferred means of loading a grease gun. In the most critical applications, you may test the grease for contaminates. There are also laboratories available that will test grease samples. If there is any doubt of contamination, have the grease tested.

If the grease gun is filled by either sucking the grease into the barrel of the gun or spooning it in, be aware of the need to avoid air pockets in the filled area. These could be introduced into the grease delivered to the bearing. Everything that comes in contact with the grease gun and the grease must be free of contamination.

Any compressed air used in cleaning the grease gun must be filtered close to the point of use. Control compressed air use so it does not stir up dust in the workspace. Once the grease gun is filled, pump grease into a receptacle until all air is purged from the delivery line or hose. If an ultrasound lubrication adapter is used or permanently attached to the grease gun, it also must be purged. Wipe the end of the female grease coupling after each use and wrap the coupling with clean plastic when not in use. Protect the grease gun from contamination by placing it in a clean and sealed plastic bag to transport it.

Grease guns can easily produce over 4000 PSI of pressure. This can damage a bearing by damaging a seal or forcing a shield into the rolling element if grease delivery is not carefully controlled."

Information from the Acoustic Lubrication Guidelines for Rolling Element Bearings in Electric Motors
Copyright 2015 Reliabilityweb.com.

Understanding the roles of maintenance vs. operations.

Operations wants to keep the equipment running until it breaks, maintenance wants to stop it at a convenient time, so it will never stop unless it is planned.

Include operations in training and decisions to schedule equipment maintenance. Teach them to understand big picture productivity. Short term vs long term thinking.

Develop a relationship between the lube tech / maintenance person and the operators. Use their eyes and ears. Set up visuals for the operators so they can understand and become part of an early warning system.

When executing a work order, survey the site upon arrival.

- Is everything labelled?
- Is the work area safe?
- Does the work order match the task?
- Are the grease fittings, seals, breathers marked and visible?
- Are pressure gauges in the green area, do they even have a green area?

Check all the fluid levels. and everything required on the work order.

After the service, update your view of how things look, sound and smell.

The site glasses can tell you if the equipment is leaking, if the fluid is clear, foamy or milky. Is the site glass painted over?

Check that the grease fittings are visible, clean, do they have color coated caps, are they painted over?

Checking out long term leaks, maybe the gear box is dry. If it has been leaking, ask who put down the spill pads, then find out how long ago and why.

Determine what is normal and safe for this operation.

Edit the work order after service is complete.

Clean up your service cart, and make sure all the tools are back in place. Clean up the cart and return the paperwork to close out the work order. Return the unused lube to inventory. Take care of the waste oil / lube, if you changed it or send out the oil samples. Return the tools and cart to the lube area.

Let your maintenance people get to know the plant by walking around. (MEBWA). Maintenance Excellence by Walking Around.

Look, listen, sniff. **But don't taste anything**. Use an instrument to see if it's hot or cold.

An often overlooked area on most equipment is the FLR. Filter, Lubricator, Regulator. These units require service, and can be put on the walk around for your operators. First keep the exterior of the units clean. Typically there is a filter unit to pull out particulate, and sometimes a water separator. Disassemble and either replace the filter element or drain the water from the separator bowl. The next unit is the regulator and gauge assembly, it is second as it should get clean air to it. This unit doesn't require a lot of maintenance other than keeping the exterior clean. If the gauge is broken or the glass is broken replace it. The third unit is the lubricator. This unit will often have an Allen screw at the top and should be set to add one to two drops of lube per minute to the equipment to keep the mechanisms operating freely. If it has a site glass, keep it clean, but make sure the bowl is filled to the line with clean machine oil. These units typically run dry from lack of attention. They do a lot to keep the critical equipment working with clean dry and lubricated air at the proper pressure.

Get trained in Oil Analysis, Thermography, Vibration and Ultrasound. There are lots of free training available from the equipment manufacturers. You can hire this equipment out or rent it.

STEP 4

FOCUSED IMPROVEMENT EVENTS

Set up an event with one of your most critical pieces of equipment. Who should be the Players in your focused improvement event? Maintenance Personnel , Operations Management, Operators, Engineering, Finance, Material handlers, getting someone from the "C" suite will add a lot of credibility to the program.

Develop the "keep it running" mindset vs the "fix it when it breaks" historic plan.

Sit back and look at these critical pieces of equipment, once you have them running, ask what it will take to keep them running? Not what do we need to keep on hand to fix it the next time it goes down.

This is important.

The focused improvement event plan is to shut it down, take it apart, clean it, then put it back together in like new condition. Put it in its most productive state.
Now the goal will be to develop **countermeasures to contamination, lubrication and access:**

These become ways to mistake proof your maintenance efforts. The goal is to be able to determine when equipment needs service with:
No outside tools, just captive wrenches, and latch handles. Opening no doors, use clear Lexan covers over belts, chains, and other drive components to view condition, lubrication and contamination. Leave access to do maintenance inspections with the least skill level required.

This example requires no tools to open and look into the drive line, also no safety issues are created to service this component.

Here is what is included in a typical focused improvement event:

- Put together a cross functional team.
- Measure and calculate OEE.
- Your goal is zero equipment stoppages.
- Measure your repetitive failures.
- Open up the equipment.
- Look for:
    - Contamination
    - Lack of Lubrication or too much lubrication

    - Air Leaks
- Stabilize the equipment
- Close up the equipment.

Duplicate:

To the next most critical machines

Pay for much of your predictive equipment with the savings generated from just air leaks. This along with less oil change time, and lower disposal fees from doing oil analysis will net big savings. Track them and use them to your advantage.

Here is an example of what a small leak can add up to over the next year. ¼" diameter leak can be $8,000.00 based on 100PSI house pressure and leaking 24 hours per day for the year. Where can you use the more than $8,000.00 you will save in your maintenance department.

With Focused Improvement we analyze, then evaluate the equipment. We use 5S principles relative to Quality, Setup reduction and mistake proofing.

- We strive to improve OEE.

- Analyze the failure history.

  Look for downtime relationships

  Look for repetitive failures

- Clean and Inspect.

- Locate maintenance issues and minor defects.

- Develop countermeasures against contamination, accessibility and lubrication.

- Enhance our preventive maintenance, faster, quicker, and more effective.

- Develop and record measurements of: Temperature, Sound, Infrared, and Air Speed to benchmark the equipment.

Establishing a walk around plan for your operators:

Start this on your critical machines, and then expand this as time permits to all of your equipment.

Here is an example of a good walk around plan developed by a company called **Tier One LLC** in Newtown Connecticut for their operators.

At the beginning of each shift they walk around their machines in a counterclockwise fashion and look at their 8 spots on the equipment to determine is it ready to run or should they call the maintenance person.

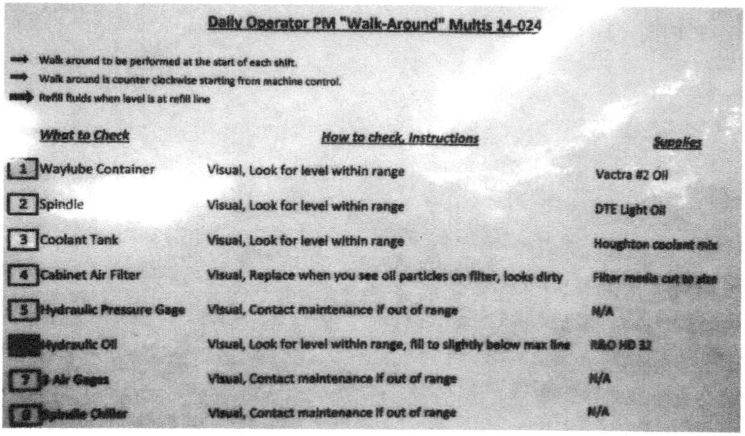

Daily Operator PM "Walk-Around" Multis 14-024

- Walk around to be performed at the start of each shift.
- Walk around is counter clockwise starting from machine control.
- Refill fluids when level is at refill line

	What to Check	How to check, Instructions	Supplies
1	Waylube Container	Visual, Look for level within range	Vactra #2 Oil
2	Spindle	Visual, Look for level within range	DTE Light Oil
3	Coolant Tank	Visual, Look for level within range	Houghton coolant mix
4	Cabinet Air Filter	Visual, Replace when you see oil particles on filter, looks dirty	Filter media cut to size
5	Hydraulic Pressure Gage	Visual, Contact maintenance if out of range	N/A
	Hydraulic Oil	Visual, Look for level within range, fill to slightly below max line	R&O HD 32
7	[illegible] Air Gages	Visual, Contact maintenance if out of range	N/A
8	Spindle Chiller	Visual, Contact maintenance if out of range	N/A

This company even has their operators add lubricant when required. The following are the stops on the operators walk around at the beginning of the shift.

Here the operator checks to see that the lubricant is between the lines, and has a matching color coded can to add lube if required.

At this spot the operator needs to make sure the lubricant is also between the lines and add if required.

At point 3 the operator checks to see if the gearbox oil is between the lines.

Here the operator is required to make sure the filter material over the electronics is clean, and replaces it when it becomes dirty. This requires taking off the magnets and putting up a clean filter. I often suggest that this material should be held on with a 1/8 inch piece of Velcro all the way around as it doesn't let in any dirt around the edges, but the magnet system is a huge improvement for the cleanliness of their equipment. Also, make sure the white side is out for proper filtering.

At area 5 the operator needs to just insure the gauge is between the marks, and call maintenance if it is not.

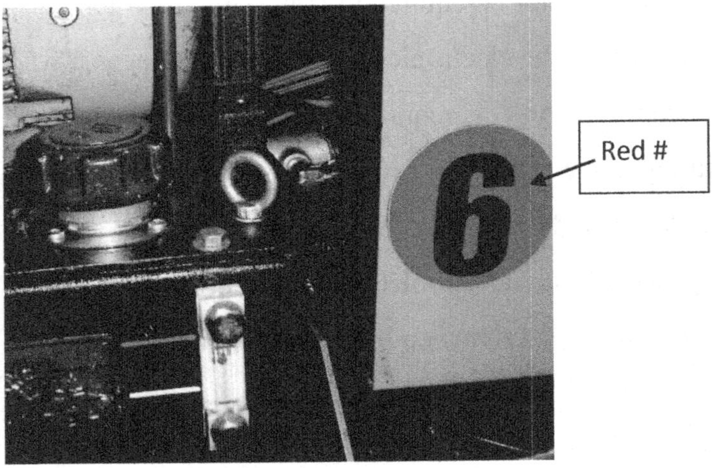

At station 6 the operator makes sure the level is within the upper and lower limits. Notice the number is red at this station vs. white on the others, this denotes using a red topped lubricant filler container, to insure the proper oil is put in.

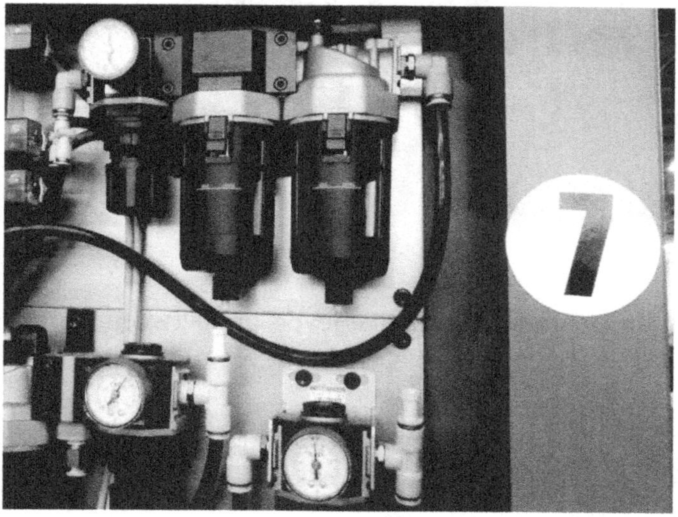

Station 7 also requires the operator to check that the needle on the gauges are between the lines.

Again the operator checks that levels are between the lines.

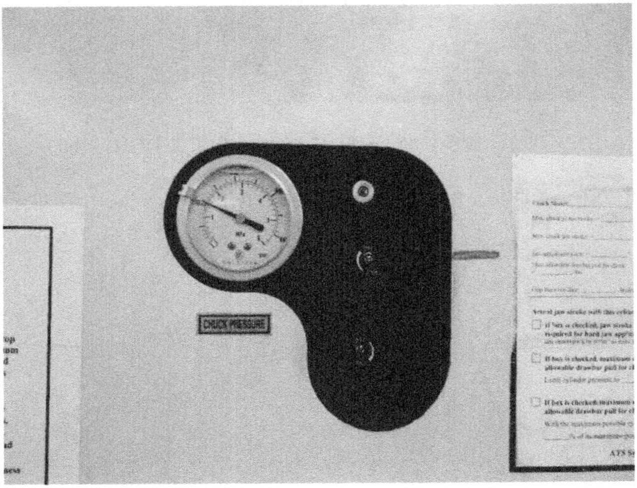

Here is an example of a walk around sheet with visuals on the sheet to help even new operators see what they are looking for.

Your company may have reasons not to have the operators add lube or adjust air pressures on the equipment to keep in between the lines. But just having someone being the maintenance persons eyes and ears is vital to keeping them doing productive and predictive maintenance work vs just breakdown work.

STEP 5
FORMAL TRAINING

90% of all present maintenance professionals are not formally trained

.

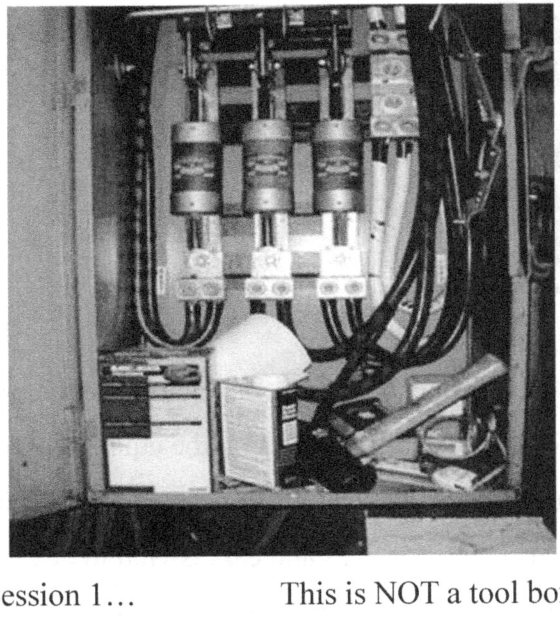

Session 1… This is NOT a tool box.

Develop a plan.

World Class Maintenance shows up directly in the Bottom Line. Fight for budget money with Dollar Based Arguments. The nuts and bolts guys must be able to talk to the brains and bucks guys to show the value of a world class maintenance program. Typically manufacturers see a 10 times increase in production to training dollar expended.

Reduce the Skills Gap

Evaluate and plan to get a training budget.

Identify Competencies and Skills that can drive your strategies and performance metrics.

- Identify the core competencies for the organization's workforce that are most critical to the organization's success.

- Use either written or verbal tests, look on the internet to benchmark your skills.
- Identify current needs as well as those for the next one to three years. Get maintenance involved with new equipment purchases. Strive for commonality of controls and an understanding of short vs. long term dollars invested.
- Determine which skills and competencies employees need, to be able to meet key goals and be effective at implementing key strategies.
- Identify the priority job skills to develop the functional competencies required for job success.
- Identify which skills and competencies are required for meeting performance standards, determine what time frame, and how they will be measured.

Assess your skills gap.

- Define the scope of how to address the gap, and consider starting with one function as a pilot.
- Use analytical tools, such as impact mapping, to identify the performance behaviors required to meet your specific maintenance goals.
- Conduct a capability audit to determine where gaps exist in employees' knowledge, skill, or behavior.

Set goals and prioritize the path to filling your skills gaps.

- Establish baseline measures of employees' current skills.
- Set goals for closing gaps between existing skill sets and those needed to support current and future goals.
- Determine which paths to take to fill particular gaps: outsourcing, hiring, training, development, coaching, and so on.

- Determine how filling skills gaps will be measured. Formal testing, On the job testing, or internet training and testing.
- Create an internal communication plan to educate and involve managers and employees in programs to close skills gaps. Train operators to become the eyes and ears for maintenance. Teach a walk around program for operators to learn what to look for and listen for and sniff for.

Implement Solutions

- With key leaders in the organization, prioritize solutions and secure funding and resources needed to do the work.
- Connect with your local Workforce Investment Board (WIB) and Department Of Labor (DOL) to identify opportunities to partner, and to access training funding that can assist your organization in addressing skills gaps. Use equipment and materials vendors for some free training opportunities.

- If possible create an organization-wide plan to address skills gaps.
- Prioritize what the organization can implement and sustain, even if there are disruptions, changes, and challenges to the organization.
- Ensure there is senior leadership buy-in and engagement in every step of implementation and communication, and that they are participating as teachers.
- For identified knowledge or skill gaps, create learning and development opportunities, using appropriate delivery options and leverage all forms of learning: formal, informal, social or lunch and learns.
- Create individual learning plans and paths for employees. Have systems in place to measure and document skills and competencies before, during, and after training and development. Link all learning and development goals to the organization's skill needs and priorities.

- Have a balance of quick wins, medium-term successes, and longer-term development so that measurement of success can begin immediately.

Monitor and measure the results and communicate the impact.

- Be prepared to demonstrate how closing the skills gap has increased organizational performance, while improving productivity and reducing costs.
- Ensure the measures are truly aligned with the organization's measures of success.
- Use assessment tools to measure progress before and after learning and development takes place.
- Identify a successful employee, work with that person to communicate the results.
- Communicate progress and results regularly to all employees.

Develop effective PMs by condition based analysis:

Consider route based oil analysis to drive condition based oil changes. There are route based infrared spectroscopy tools that allow your lubrication technician to follow a planned route testing a few drops at each sample point to obtain the Total Acid Numbers (TAN) of the oil, water content and oxidation levels in just one minute. Consider route based vibration analysis. Get your lube tech to start with a pilot area to do a lube audit. Send this to a lab of your choice, develop a sampling route and timing, then put this into the work order system. If you have one start doing it, keep adding more equipment to each route by criticality. Train your lubrication technician to use ultra sound to properly grease bearings on their route. Use this data to catch bearings before they become a breakdown.

If you are currently in breakdown mode free up time to get more time for your condition based PM's. Start teaching the steps of trouble shooting to get out of breakdowns sooner.

Develop the" keep it running" mindset, with training. What things can you think up to catch breakdowns before they happen?

Consider an Interim Maintenance Manager. This is a good option if your head guy is retiring, not only can an interim maintenance manager keep things going, and develop some new programs, they can help interview for the new replacement.

Out Source Maintenance Duties.
Effective PM's can be done by the equipment manufacturers, and they can help you develop a program your people can follow. There are many companies offering this service on a contract basis.

Who in the Company stands to gain if Maintenance is not invested in. Short term the bottom line looks better but long term total maintenance costs will skyrocket. Learn to sell the value of investing in reliability and maintenance.

We all know the costs of poor reliability are huge.

Reliability Centered Maintenance (RCM) looks at identifying what should be done to an asset, to assure its functions in a safe and cost effective way, by asking 7 questions.

1. What are the functions and performance associated with the asset in its current operating condition?
2. In what ways does it not perform to its full potential?
3. What is the root cause stopping it from performing its full potential?
4. What happens when each failure occurs, concerning quality, productivity and customer satisfaction?
5. What is the impact of each failure, in time and dollars?
6. What can be done to predict or prevent each of these failures?

7. What should be done if adequate work could

not be performed before the failure?

Learn the techniques of Troubleshooting.

Training like this can get you more time to do proper PM's.

There are some steps to free up time to do condition based PM's.

Typical troubleshooting takes a lot of time, so do it properly and start saving time. As you know wasted time is wasted money.

Prepare before jumping right into a downed machine.
What should the equipment be doing?
Understand the sequence of Operations.
Do you have the right skills in place?
Have you Established a safe work environment, LOTO?
Gather information on what happened prior to the failure.
Talk with the equipment Operator.
Use the 5 "Whys". Are there diagnostics on the

machine? Do a root cause analysis, then correct it.

Prevent a reoccurrence with:

- TPM
- Predictive tools
- Reliability Centered Maintenance.

All maintenance organizations need to minimize waste to keep their equipment running and improve profitability. The ability to troubleshoot equipment quickly, effectively and permanently will keep these failures from returning, and this in turn grows OEE and will improve your bottom line.

Train and empower your operators to improve reliability. No one knows the equipment better than your operators. Give them the tools to be the eyes and ears of maintenance. Let them send in a service ticket or work order form.

Give them the visuals to understand when

something is going out of spec.

Develop standard work in combination with teamwork.

PUTTING IT ALL TOGETHER

The Big Picture

Things that must be addressed to get to world class.

Identify the Critical Equipment. Pick your top three pieces of equipment that will hurt the most in terms of production or cost to repair, or whatever is the most detrimental to the business if it goes down. Then pick the next two pieces of equipment that are the biggest pain in the butt in terms of taking your time and effort. Look for a way to upgrade them or consolidate them into another process or machine.

Collect Data Identify a cost per machine for maintenance and then OEE in order of criticality. With this data attack the 6 major losses starting with idling and minor stoppages. Set target BOB's for OEE. At every breakdown look for the root cause. Was it due to contamination or lack of lubrication? It probably was.

Control your spare parts and tools. Identify what you have, 5S the spare parts area so you can get rid of what you no longer need and can find what is left. Determine what must be stocked, 24 hour rule, and mission critical parts.
Develop a PM plan for each of the critical machines in order. Develop the PM from the manuals, and the OEE data on what is causing breakdowns. Put countermeasures in place to guard against contamination and lubrication and access issues.

Predict failure before is occurs.
Start using predictive tools, like Ultrasound for bearing anomalies and air leaks. Infrared for motor temperature issues and alignment. Find loose and potentially dangerous issues in your electrical cabinets. Vibration analysis for rotating members,

and oil analysis to see what issues are lurking in your gear boxes.

Get out of the breakdown mode.

Start to change your mind set. Instead of how do we fix this as quickly as possible, start to look at your most critical and troublesome pieces of equipment and think about "What do we do to keep equipment running".

Look at mistake proofing specifically how to keep the equipment clean, well lubricated, and how to gain easy access to major subcomponents, for condition based PM's. A simple example is filtering your motors. Start by cleaning them. Put a 1/8 inch piece of Velcro all the way around the fan and put on the filter material white side out.

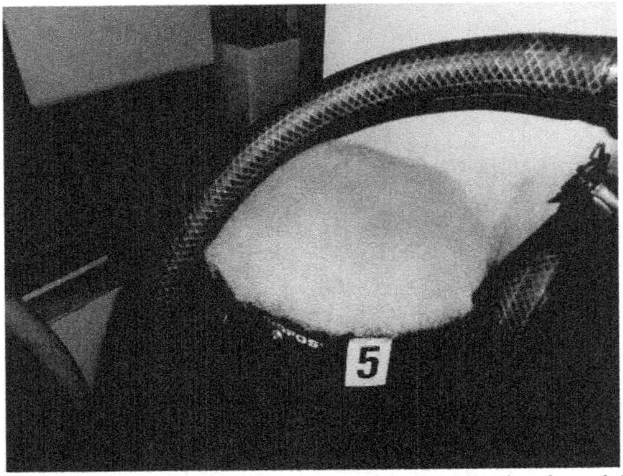

If this material had a white side it should be put toward the air flow.

Develop route based oil analysis to drive condition based oil changes.

Address your Skills Gap Even if it is one person, what skills do they have and what are needed. Find out by interview, or online testing, look for help from state agencies to fund training, or many of your equipment or supply vendors will give your people training for free.

Where are you on the PF curve?

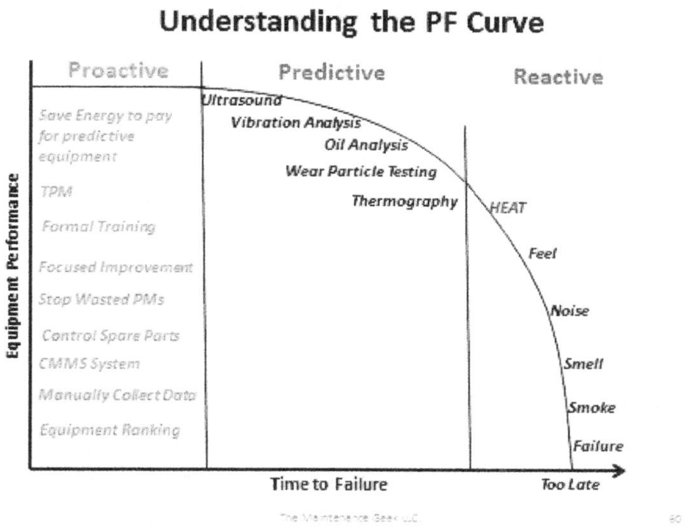

Understand the P-F Curve.

Reactive mode putting out fires, only doing repairs when the equipment is broken.

Predictive mode, using some predictive tools, like oil analysis, vibration analysis, or thermography.

Proactive mode. You have ranked your equipment, started collecting data and recorded it somewhere. You actually look at it and use it.

Put it in a CMMS system or at least use an Excel spread sheet for each machine and save it by asset number. Schedule your PMs with Outlook and link the Excel sheet to it for each machine to track PMs. E mail these outlook tasks to your maintenance people and Outlook will track schedules and completion dates. Get control of your spare parts, institute better PMs that cause less damage than they incur. Start a focused improvement event on your most critical equipment, and get training where needed.

Start saving time troubleshooting and mistake proofing your critical equipment, to free up time for condition based PM's.

There are some steps to develop a simple PM form in Excel to track your PMs.

Asset #	Description	Items to service	Scheduled	Actual Date	Parts required	Actions Taken	Date Comple

Start with your asset number and description, you will want to name these by asset number where ever you plan to store them. The biggest section will be "**Items to service",** I will show you lots of them, you only need to delete the ones that do not apply. You can customize the rest of the columns for schedule dates and actual dates including notes.

Here are some things to look for on a PM,
under the items to service column.

Items to service

Items from the Manual

Is the equipment level?
Soft foot?
What is unique to this equipment?
Structural Bolts and Frames

This part of a PM may only have to be done every couple of years. But the basics need to be looked at.

Mechanical Systems

Rolls and Rollers

Conveyors and Traction Belts

Shock Absorbers
Pillow Block / Flanged Bearings
Transport Chains and Drives
Hand Crank Assemblies
Guide Rails

On the mechanical systems make sure you put plenty of action items in the PM to reduce the occurrence of pencil whipping.
For example rolls and rollers; Inspect roll surface, verify roll turns freely without runout. Inspect bracket for damage, insure brackets move freely, align rollers to the rolling surface and parallel to each other. Lubricate axle with light oil.

Conveyors and traction belts; Inspect for wear rips and damage. Align belt tracking. Inspect condition of take ups. Replace worn take up hardware. Inspect condition of head tail and idler pulleys for damage. Surfaces need to be smooth and concentric without gouges.

Pneumatic Systems

Cylinders and Actuators
Filter / Regulators /Lubricators
Valves and Solenoids

An example for Valves and solenoids.
Test fire valves and solenoids electrically and with mechanical override. Tighten all mounting bolts, inspect exhaust ports and change any dirty mufflers. Inspect for and repair any leaks. Perform a runtime inspection for slamming. Adjust flow controls on solenoids and cylinders.

For FLR service; Drain filter /regulator / lubricator. Clean FLR bowls and filter elements in soapy water.
Insure O Rings are installed properly, grease before reinstalling. Verify seal after replacement. Inspect manifold for leaks and repair. Fill oiler with correct air mist oil. While operating verify drip rate of oiler. Typically 1 – 2 drops per minute, check manual. Set pressure on regulators to the manual specs.

Hydraulic systems

Pumps
Tubing Hoses
Piping
Valves
Tanks
Filters

Motor / Drives

Motors
Gear Reducers
Chain Drives
Indexing Clutches
Synchronous Belt Drives
V and Transport Belt Drives
Lead Screws / Ball Screws
Clutches
Couplings
Open Gears
Universal Joints

Controllers / PLCs

Disconnects
Transformers
Junction Boxes

Lubrication Systems

Lubricate and inspect. See FLR suggestions.

Electrical Systems

Control Panels: Pressure or Vacuum?
Proximity and Limit Switches
Blowers
Encoders and Resolvers

Safety / Guarding
Photo eyes
Pull cord safety switches

Tooling Fixtures
Suction cups
Vacuum Generators
Linkages
Bushings
Springs
Bumpers and Hard stops

Take out the items that do not apply and save the sheet for each machine by asset number.
Use Outlook to schedule your PMs by Yearly, Monthly and Weekly PMs. Color code the classifications and email the tasks to your maintenance people. Outlook will keep track of the tasks and let you know when they are being done. Attach the Excel sheet to the outlook task and as they update it, they save it to the asset number established. This Excel / Outlook system will work for one or two maintenance people and a dozen or

so pieces of equipment but if you get bigger really look at a CMMS program. Many will let you test drive their system for 30 days to see if it adds value to your organization.

Get Money allocated: 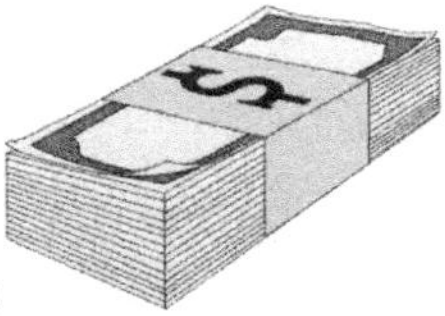

Work on convincing the corner office. Speak their language of dollars and cents to get funds available to update critical equipment, $1 dollar spent on preventative and predictive maintenance can translate to 5-10 dollars spend in breakdown mode. What does it cost you right now? If you are a $2 Million to a $10 Million dollar company with lots of equipment it typically costs $2,000 - $10,000 per month for maintenance costs. That is $24,000 - $120,000 per year. Allocate $2,000 - $10,000 to get out of the breakdown mode. Go back with proof, and do it again next budget year. In 5 years you will be out of the breakdown mode, if you do it right. Your goal should be zero unpredicted failures.

Fix your air leaks. $3,000.00 spent on an ultrasonic leak detector and some overtime fixing leaks in order of size will return 10 times that number. Keep track of what you save and get a portion put back into your budget. Consider the intelligent use of compressed air for efficient blow offs, cleaning, drying and eliminating static, and possibly electrical cabinet cooling.

Update critical equipment, buy mission critical spares, then purchase and train on predictive equipment. Understand and train on all your tools.

Use maintenance techniques to save money. Document ampere draw improvements, add pictures and record. Utility companies tell us that every dollar saved in the operations by a manufacturer could be worth $7 - $12 in sales. Track your savings and let management know that it is the equivalent of sales having to come up with 7 to 12 times that amount added directly to the bottom line. That they understand.

Departments vs. Cells:
As we use more lean techniques in our facilities to approach one piece flow, this is a move toward flow vs discrete production, the consequences of downtime is magnified, as one machine goes down in the cell everything stops. Conduct some focused improvement events, once procedures are set up and operators have work order forms, pick a representative machine or critical machine. Conduct a TPM event, write a PM schedule based on the PF curve.

In the United States, we spend more than $180 billion every year on wasteful maintenance related practices. Travel time, waiting at the parts crib, looking for maintenance tools, temporary repairs, & emergency shipments of repair parts, hunting for critical parts, are just a few of those wasted practices. Consider the time spent wading through parts you do not even have machines for anymore, buried parts, damaged parts, and lost time due to inaccuracies in your spare parts inventory.

It is typical that out of an 8 hour day a Maintenance technician only gets about 2 hours of "wrench time" when they are actually adding value.

Use the Value Stream Mapping tools you already understand from your lean training, to map your maintenance efforts. Although there is some additional training required you have most of the tools already to pick up some great gains in wrench time for your maintenance personnel. Think about Value Stream Mapping from a Maintenance perspective.
Bring in a maintenance consultant on a short-term contract. The consultant can create a maintenance plan and train the company's maintenance staff to implement that plan. The advantage of this approach is that companies can typically gain

access to much more experienced maintenance professionals at a more affordable cost. However, it's merely a short-term solution since that consultant will be moving on at the conclusion of the project. The long-term success of this approach rests on the ability of the plant's leadership to implement the program after the consultant has left.

Companies that treat maintenance as a priority can save millions of dollars by avoiding expensive equipment repair, productivity stumbles, and safety-related losses. It's not difficult to make maintenance an important part of day-to-day operations, but it does often require company leaders to change the way they think. Those companies that do make a commitment to maintenance don't just save money—they also create a better work environment, produce a better product, and gain a competitive edge over their competitors who don't understand the importance of maintenance.

Outsource, short term:
Consider hiring an Interim Maintenance Manager. Bring in help along the PF curve to supplement while you develop the right skills.

Equipment Manufacturers for PM's
Check the owner's manuals for a place to start. Start using condition based vs time based to improve your PM's.

Outsourcing Maintenance Duties
Have an ultra sound company come in and tag all your leaks . Then hire out or have one of your people fix the leaks, even on overtime it is money well spent. Do not pay to have them tagged, if you are not going to fix the leaks. There is funding money available from most Utilities for finding and fixing wasted compressed air.
Call in an infrared company to do an infrared survey if you can't afford your own unit $1000-

$10,000, then fix what they uncover.
If you already have any of the predictive tools use them around your critical equipment to find air leaks, dry bearings, and loose connections in electrical cabinets. Many maintenance organizations have these tools but no one has been trained. Get some training; it is cheap compared to the returns. Use these tools on a regular schedule, fix what is found, put in countermeasures for contamination and lubrication then watch your incidents of breakdowns drop way off. Get lube reps involved in setting up a good lube program. It is usually free. By learning good lube practices like cleaning fittings and lubing new chains you can start to see a 10X life factor on these items.

This will give you more time to do condition based PMs. By putting this all together getting out of the breakdown mode can be good for your health, in addition to your equipment's health and your company's health. Less stress for you and a bigger bottom line for your company.

Move from Reactive to Predictive and then toward Proactive tools and thought processes.

This is how to get out of the breakdown mode.

The Maintenance Geek.

www.maintenancegeek .com

Follow the Maintenance Geek on the web site, or call at **(203) 510 - 8375** to talk about how you can get out of the break down mode. We have done many programs with companies with only one maintenance person, it is a pure joy to see how much more productive companies can become when they use a few simple tools.

Best of luck in your journey to a world class maintenance organization, let me know how you are doing.

ABOUT THE AUTHOR

Robert Kravontka is a Maintenance Geek, who was tired of living break down to break down. He has studied the thought leaders in the industry, and is bringing that knowledge to his followers at

www.maintenancegeek.com

Robert lives with his wife Sharon in Waterbury Connecticut, he has an engineering degree and an MBA. His hobbies include running his car, and heating his house with vegetable oil, and has enjoyed building an electric car.

Made in the USA
Monee, IL
11 November 2026

68342184R00069